Use It

Or

Lose It

Boost Your Sexual Energy

Violet Karma

COBY BLAZE
DESIGNS

Violet Karma
©2019, All rights reserved.

ISBN-13: 9781987081985
ISBN-10: 1986167879

Notice

This book is intended as a reference volume only, not as a
medical manual.
The information given here is designed to help you make
informed decisions about your health. It is not intended as a
substitute for any treatment that may have been prescribed by
your doctor. If you suspect that you have a medical problem,
we urge you to seek competent medical help.

Table of Contents

Introduction

No matter how much you and your partner enjoy the old favorites, you might want to have some fun now and then and experiment with something totally new. It's true that there is only a handful of core positions, but each of these have almost infinite variations. Use It Or Lose It is designed to help you increase blood flow to vital sexual organs through light, healthy exercise and eating. The less you use your sexual muscles the weaker they get thus "Use It Or Lose It". Your sexual relationship may be used to the common man-on-top, woman-on-top and rear-entry. Being petty with variations may cost your relationship or worse, without any workout at-all to keep fit-for-sex. This book contains easy workouts with massive sexual gains, explained in details of each workout plan. We do not recommend organ enlargement by quick-fix programs flooding the markets but vitamins and minerals that create a natural boost to go with the workout plans are recommended. (Explained to detail).

"Exercise for sex and use sex for exercise"

A woman can have sex whenever she chooses. True, it might take her longer to get ready as she gets older, and she might get less wet than she used to, but she can always pull it off in a pinch. Men, well, have to stand to deliver. There's no faking it. After a long and stressful day — and, let's face it, that often means every day — you might not have the physical stamina to achieve a long-lasting erection or the energy for the subsequent lovemaking. This is more likely to occur when you're out of shape, but even the fittest men can benefit from a little extra help. Why not take advantage of some "magic" root, berry, fungus, paste, or pill?

"Whether you think you can or cannot, you are right"

Obviously, you don't want to believe every advertisement you read — and the word magic, which appears on labels with amazing frequency, should be a full-throated warning that you're about to throw away your money. Manufacturers have been quick to exploit the natural male desire for extra sexual oomph. If you believe that certain herbal concoctions will give you sufficient animal magnetism to lure women away from an absorbing manicure and into your bed, well, there are plenty of charlatans ready to take your money.

"Low libido accounts for roughly half of all admissions to sex therapy clinics."

You need to be healthy to have good sex — but frequent sex can also make you healthy. Animal species that behave promiscuously tend to have higher levels of white blood cells and greater resistance to infection. In human studies, those who have sex twice or more weekly have higher levels of immunoglobulin A, a disease fighting antibody. They're also half as likely to have a heart attack or stroke. If you have a partner who is open enough to let you that your sex life needs help then you should be glade, most couples don't and in turn destroys your relationship. Sex shouldn't feel like a house chore like most marriages today, nor does it have to be one-sided. No!

WORKOUT

WOMEN APPRECIATE MEN WITH GOOD PHYSIQUES JUST AS MUCH AS M EN ENJOY FIT WOMEN. IMAGINE THAT.

Men will always be more visually inclined than women, and you can thank your lucky stars for that. Very few of the women we talked to listed Stallone-like muscles at the top of their wish lists. In fact, they're far less likely than men to reject a potential suitor because he's in less-than-perfect shape. But do they appreciate a well-toned body? You bet.
"A muscular upper body and strong arms are a real turn-on," said a 41-year-old hospital administrator in Buffalo. "I love toned legs, when you can see the outline of his calf and quad muscles," said a 33-year-old Portland, Oregon, biologist. And this from a 37- year-old medical student in Albuquerque: "Some men have a great chest or a nice butt that you just want to get your hands on. But that doesn't mean he has to be 6 feet tall or as strong as an ox."

If nothing else, it's worth staying in shape — or getting there if you haven't yet made the commitment to buckle down in the gym — just to make yourself more attractive to a current or future partner. The benefits don't stop there. Men who stay in shape have more energy.
More energy for sex, for working, for life in general. Exercise makes you happier and more self-assured. Most important, it improves the machinery — blood flow, nerve activity, and brain chemistry, for example — that you need for full-bore sex appeal and performance.

Fit For Love

How strong do you have to be for sex? Stronger than you might think — unless your idea of passionate coupling is to imitate one of the stiffs on Six Feet Under. Even when you're in the passive, man-on- the-bottom position, you need strong abs and hips to push your body upward. Men who are really out of shape often find themselves getting fatigued before the act is done — as well as sore the next day. That's not very sexy. Or satisfying. And that's an easy position.
How are you going to feel when you try something more athletic?
If you plan on doing anything more vigorous than just lying there during sex, you're going to need strength, flexibility, and endurance.

"Life is good, sex should be better"

Consider the man-on-top position. You need a strong chest and strong arms to move with any kind of freedom as well as to hold your weight off your partner. Sex while standing? You'd better have some muscles in your calves, hamstrings, and butt.
"When I'm on top, I find it hard to hold myself up for long periods of time," admits a 21-year-old student in Pittsburgh. "With all the thrusting and whatnot, my back and abs seem to give out sooner than I'd like." The doctors we talked to said that this guy's story is hardly unique. Even though men are more health conscious today than they've ever been, that doesn't always translate into committed workouts. Basically, we spend way too much time at our desks, sitting in our cars, and taking up couch space. Our sex lives are paying the price. Even our desire for sex may be heading downhill.

It's not news to exercise physiologists that men who are sedentary can have significant drops in testosterone, the "male" hormone that fuels libido. Add to that the natural drop in testosterone that occurs when men get older, and you can see why a lot of us aren't exactly thrilled with our bedroom oomph.

None of this is inevitable. The amount of exercise that you need for optimal sexual fitness is modest. Half an hour or so in the gym most days of the week will give you endurance, energy, and sex drive that you didn't even know you had. Harder workouts, if you're so inclined, can boost your performance even more. Finnish researchers recently reported that 30-year-old men who engaged in regular weight lifting had significant boosts in testosterone. More testosterone quickly translates to more muscle bulk, faster recovery after workouts, and in some cases, a boost in libido.

"I'm living proof," a 23-year-old student in Boston told us. "I lost 80 pounds and started lifting. I now have more sex and better sex for longer periods, with no next-day soreness." Better sex by itself should be reason enough to strive for the exercise edge. But it's not the only reason.

-Men who lift weights get more limber, not less. Your balance will improve as well.
-Strength training increases bone strength and density. That means less risk of fractures.
-It "oils" the joints and lowers the risk of arthritis. It makes it easier to ski, run, or just get around in this world without bum joints.
-It increases the level of beneficial HDL cholesterol, which carts the artery-clogging stuff out of your bloodstream before it lays down rock-hard deposits in the arteries that supply the penis.
-Men who start a strength-training program invariably

report a huge increase in confidence — the quality that nearly all women rate at the top of their sex-appeal charts.

-Whether you're interested mainly in building muscle, strengthening bones, or increasing sexual stamina, lifting weights should be your first choice. Muscles grow only in response to high-intensity overload — pumping iron, in other words. Moving your arms up and.

Sex Workouts

Stretching, strength training, and cardiovascular workouts are the guts of the built for sex plan. Exercise increases the flow of energizing endorphins and adrenaline, body chemicals that make you feel strong, sexy, and confident. Guys who work out tend to have higher levels of testosterone, the male hormone you need for arousal as well as performance. Researchers at Cologne University Medical Centre in Germany even found that men who exercise have substantial increases of bloodflow to the penis — bloodflow that you need for erections.

We're not talking hard-core athletics. Men who work out moderately have increased sex drive and more predictable erections. In a landmark study, 78 sedentary, middle-age men did nothing more than aerobics four times a week. When researchers talked to them
9 months later, they learned that their frequency of sex had increased by nearly a third, and the frequency of orgasms had increased 26 percent.
And there are other exercise payoffs.

INCREASED CONFIDE NCE.

Men who exercise feel stronger and more energetic. They have more physical and mental confidence. They feel good in their own skins, and this goes a long way in the sexual realm.

MORE ENERGY.

Even moderate levels of exercise—lifting a couple of times a week, for example, or taking daily walks—can stave off sex-killing fatigue. As a 54-year-old technical writer from Phoenix told us, "I was sucking down pastries, chips, and sodas, and I was tired all the time. About a year ago, I made up my mind to get back to exercising, eating the right foods, and walking at lunch. I think my hormone levels must be up, because I want sex all the time."

LOOKING BETTER.

Lift weights a couple of times a week for a few months, and you're going to see the difference. You'll feel better when you look in the mirror. Your partner will enjoy looking at you. Now that's sexy.

BETTER S EX.

The more you exercise, the more blood arrives in the penis. More blood means better erections. You'll be able to go longer without fatigue. And you'll want sex more than you did before.
You'll also be stronger and more limber—qualities you'll need if you even hope to try some of the more creative positions out there.
Whether you want to get creative or stick with the tried and true, one simple exercise can kick it up a notch. Look at the "Position of
Strength" section in chapter 8 for the specific exercise that strengthens the muscles you use for your favorite positions.

LESS STRESS.

It's probably the main factor in low libido as well as erection problems. In fact, men who are stressed all the time find it almost impossible to have good sex because of destructive changes in body chemistry. Research has clearly shown that even lightweight exercise programs, such as occasional lifting, stretching, or aerobics, cause a dramatic drop in stress chemicals and an increase in sexual interest and activities.

BETTER HEALTH.

It's hardly a secret that men who exercise are a lot less likely to get diabetes or hypertension, chronic and life-threatening diseases that are among the main causes of sexual problems. Because exercise is so effective at improving appearance and physical and mental health, we've designed a total fitness plan that includes strength training, stretching, and cardiovascular workouts.

You'll get stronger, in and out of bed. You'll have more sexual endurance, and you'll notice improvements in the muscles and joints that you need for good sex. We've even included exercises to improve ejaculatory control and increase the intensity of orgasms.

Positions Of Strength

After a dizzying 54 pages of workouts, how can there be more? The next 13 exercises are specifically targeted to sexual positions mentioned in this book and are the most effective workout for the position they're paired with. For example, the missionary position works best if you support your weight over your partner's body, leaving some space in between. What bears the bulk of that effort? Your shoulders — and the best exercise to strengthen them is the lateral raise.

The positions we chose range from everyday basic to spice-it-up complex. We picked perennial favorites (missionary, woman on top) as well as ones we think you'll like even if you don't try them more than once (the Honeybee, the Pretzel). You can incorporate each exercise into the workouts described in this book by prioritizing one or two each day at the beginning of your workout. Or, if you're not up for the full exercise program for whatever reason, pick the exercise that supports your favorite position and do that one only. That's an ideal strategy for when you don't have a lot of time to work out or you're on the road or under the weather. Even if you can't get a full workout in, your sex life won't suffer.

Position #1: Man On Top

It's probably the most popular sexual position in this country because it allows face-to-face intimacy and full-body contact and permits deep thrusts and full penetration. It's comfortable for the woman because she can lie on her back, with or without a pillow to change the angle; men like it because they can control the timing and penetration.

There are a few drawbacks, however. This "missionary position" can be too arousing for men in some cases, causing too-quick ejaculation.

A small woman with a bulky partner may find his weight uncomfortable unless he takes care to support his weight on his arms. From a man's point of view, it can be tiring because he's doing most of the heavy lifting.

Man On Top (Variation)

The woman can wrap her legs around your waist or neck or rest them on your shoulders. You can raise her legs by putting your forearms under her knees.

Man On Top (Variation)

The woman can pull her knees to her chest and put her feet flat against your chest.

Why she likes it: Deep penetration, ease of kissing, body-to-body closeness. The so-called missionary position hardly indicates a lack of creativity. Surveys show it's a perennial favorite of men as well as women. In this position, you also can press your pubic bone against her clitoris as you rock and thrust, giving her extreme pleasure.

Muscles to work: Shoulders. They keep your torso upright and keep your full weight off her body. Also, your naked chest and abs give her extra visual pleasure.

Best Workout: Standing Lateral Raises. Stand holding a pair of dumbbells at your sides with an overhand grip, your elbows slightly bent. Bend slightly forward at the hips, keeping your lower back in its naturally arched position. Raise your arms up and out to the sides until they're parallel with the floor, keeping the same bend in your elbows. Pause, then slowly return to the starting position. Try for three sets of eight repetitions each.

Position #2: Woman On Top

It's second in popularity only to the missionary position. The woman straddles your pelvis and raises and lowers herself on your penis.

Like the man-on-top position, it allows you to kiss and make eye contact. The woman controls the pace as well as the angle of penetration, and she can stimulate her clitoris while you're making love.

Most men find this position visually as well as physically stimulating.

You can watch your partner as she moves up and down, and your hands are free to caress her face, breasts, or buttocks. The depth of penetration equals that of the man-on-top position, which is pleasurable for both partners.

The drawbacks are the reverse of man on top. In this position, the woman rather than the man may find herself getting tired. Also, women who are self-conscious about their bodies may be somewhat uncomfortable with having them in full view.

Woman On Top (Variation)

The woman can lean forward to rest her hands on the bed in front of her or on your chest.

Woman On Top (Variation)

The woman can turn around so she's facing your feet.

Why she likes it: She controls the pace and movements.

Muscles to work: Abdominals. They provide good thrusting control and force—important when you're in the passive position, without the leverage you have on top.

Best workout: Crunches. Lie on your back, with your knees bent at 90 degrees or resting on a bench. Hold your hands behind your ears. Use your abs to curl your torso upward 4 to 6 inches, keeping your lower back firmly pressed to the floor. Pause, then slowly lower. Try to start with three sets of 10 crunches. As your endurance increases, increase it to five sets of 20.

Woman On Top (Variation)

The woman can turn around so she's facing your feet.

Why she likes it: There's deep penetration, she controls the pace, and she can stimulate her clitoris manually. In this classic Eastern position, you lie on your back as the woman straddles your penis, her bottom toward your face.

Muscle to work: Hips, abdominals, and thighs. You'll need strength in these areas to rock forward and back and to angle your hips for optimal penetration.

Best workout: Hanging Leg Raises. Using a chinup bar and an overhand grip, hang with your arms straight and shoulder-width apart. Use your abdominal muscles to raise your legs until they're a bit higher than your hips. Hold it for 2 seconds, then return to the starting position. Try for three sets of eight repetitions.

Position #3: Tabletop Sex

You can stand at the end of a bed or table and enter her while she reclines on her back, resting her legs on your chest or shoulders.

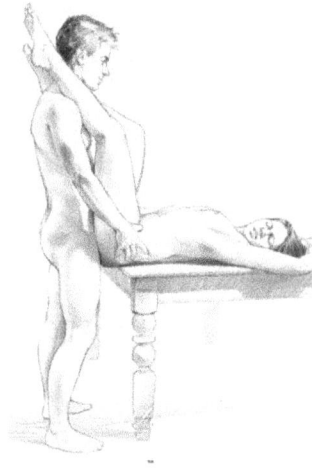

Why she likes it: Spontaneous passion, deep penetration.

Muscles to work: Hamstrings. These muscles have to be strong to have sex while standing, especially when you're bent over and thrusting forward.

Best workout: Leg Curls. Lie facedown on a leg-curl machine. With your hips flat against the bench and your abdominal muscles tight, curl your legs behind you until your feet are about perpendicular to the bench. Pause, then lower your legs slowly to the starting position, stopping just before your knees are straight. Try for three sets of eight repetitions.

Position #4: Sitting

You sit on the bed while your partner sits on your lap in a face-to-face position. It's a very intimate position that permits a lot of kissing and touching. It's visually stimulating because you can fully see what your partner is doing.
Having sex while sitting is physically taxing. You have to have strong abdominal muscles to hold yourself upright. Penetration can be fairly deep, but you can't move quickly in this position.

Sitting (Variation)

The woman can turn so that her back is against your chest.

Sitting (Variation)

The woman can straddle you face to face in a chair, then lean all the way back to rest her head on a pillow on the floor.

Why she likes it: Sex while sitting offers the best of all possible worlds. You can penetrate as deeply as you can in lying positions, with a bonus: The woman gets the opportunity to set the pace, moving as quickly or slowly as she likes.

Muscles to work: Back and abdominals. Strength in these areas makes it easier to hold yourself and your partner upright without fatigue.

Best workout: Wheel of Torture. Kneel on the floor as though doing a modified pushup, with your hands resting on a barbell. Glide forward as you roll the bar in front of you, flexing your shoulders and extending your spine. Keep going until your hips are in line with your torso. Then slide backward until you're back in the kneeling position. Try for three sets of eight repetitions.

"Sex is like snow: You never know how many inches you're going to get or how long it will last."

Position #5: Standing (From The Back)

It's a good position for spontaneous lovemaking — quickies, in other words. You can enter the woman from the rear while she leans forward slightly.

This position requires more strength and dexterity than most other sexual positions. Don't try it if you have a bad back, or if your partner is too heavy to maneuver safely.

Why she likes it: Very deep, fast penetration.

Muscles to work: Forearms, back, and biceps. You need to hold your partner firmly to keep from thrusting her into a nosedive.

Best workout: Wide-Stance Romanian Deadlift. Stand with your legs hip-width apart and your knees slightly bent. Grab a barbell with an overhand grip and your hands just beyond shoulder-width apart. Hold the bar at arm's length at mid-thigh level with your shoulders back and chest out. Keeping your back flat and your knees slightly bent, bend forward at the hips, keeping the bar close to your thighs. Lower the bar toward the floor, going as far as you comfortably can. Slowly return to the starting position, keeping your back straight throughout the exercise. Try for three sets of eight repetitions.

Position #6: Skin The Cat

In this inverted position, your partner lies flat on her back; you kneel between her knees and lift her legs up and over your shoulders, so only her head and shoulders remain on the bed, and she is suspended with her calves resting on either side of your neck. You then slide your penis downward into her.

You can grasp her thighs and pull her toward you rhythmically to assist in thrusting. You also can slide your hands down and caress her clitoris for added pleasure.

Though thrusting is easy from the kneeling position, men with steeply angled erections may find the downward motion of this position difficult.

Why she likes it: Being upside down gives her a giddy head rush and feeling of weightlessness during intercourse and allows her to caress her breasts.

Why she likes it: The giddy head-rush.

Muscles to work: Deep abdominals. When you're on your knees, and your partner straddles you in an upside-down position, you need strong abdominals to hold yourself upright and to move back and forth.

Best workout: Kneeling Vacuum. Get down on your hands and knees, keeping your back flat. Take a deep breath, allowing your belly to push out. Then forcibly exhale and round your back as you lift your navel up toward your spine, contracting your pelvic muscles. When you can no longer exhale, keep your back rounded. Hold the contraction for anywhere from 10 to 60 seconds, breathing regularly the whole time. Rest for a minute, then repeat up to 10 times.

Position #7: The Honeybee

This part-sitting, part-reclining position allows maximum penetration and visual stimulation, though it requires a fair amount of strength and flexibility from both of you. You sit, leaning back with your hands on the bed behind your back for support. She straddles your penis, placing both legs over your shoulders and holding on to either side of your neck with her hands. To thrust, you press your thighs together, lifting her up, then open them wide, letting her slide back down, repeating so your legs make a "fluttering" motion.

The tricky part is maintaining insertion while getting into position.

If she's fairly flexible, you may want to let her straddle you in the classic woman-on-top position first, then shift to place her legs over your shoulders. Otherwise, assume the position without inserting; then you can lean back and gently guide your penis into her.

Why she likes it: Her vaginal wall is angled upward, allowing for more intense G-spot stimulation with every thrust.

Why she likes it: Intense G-spot stimulation with every thrust. This position allows maximum penetration and visual stimulation.

Muscles to work: Legs, hips, and butt
Best workout: Dumbbell Lunge. Grab a dumbbell in each hand, your palms facing your body, and stand with your feet hip-width apart. Keeping your back straight, take a long step forward with your right leg. Bend your leg until your right thigh is parallel with the floor. Your left leg should be extended, with your knee slightly bent and almost touching the floor. Keep your right foot stationary as you straighten your right leg. Switch legs and repeat on the other side.

Position #8: Standing (From The Front)

You can lift the woman with your arms, facing you, and hold her thighs while moving her up and down.

Why she likes it: Full body-to-body contact, the feel of your hands on her butt.

Muscles to work: Calves, hamstrings, and butt. They work together to provide the strength needed to hold your partner off the floor.

Best workout: Leg Presses, which work all three muscle groups together. Sit on a leg-press machine with your back against the pad and your feet shoulder-width apart on the foot plate. Adjust the seat so your knees are bent slightly more than

90 degrees. Push the weight until your knees are almost locked, then slowly return to the starting position. Try for three sets of eight repetitions.

Position #9: Rear Entry

The woman gets on her knees and elbows while you enter her from behind. Many women say that this is the best position for stimulating the G-spot. It also gives her visual imagination full play; when you're out of sight, she can fill her mind with whatever fantasies she likes.

Men often enjoy this position because it gives them a sense of power and dominance, while at the same time allowing deep and vigorous thrusting. They can also reach around and touch the woman's breasts or clitoris or enjoy it while she touches herself.

The rear-entry position is less intimate than other positions, however, and some women don't like to be "taken" in the position disparagingly known as doggy style.

Rear Entry (Variation)

The woman can lie on her stomach with her bottom elevated.

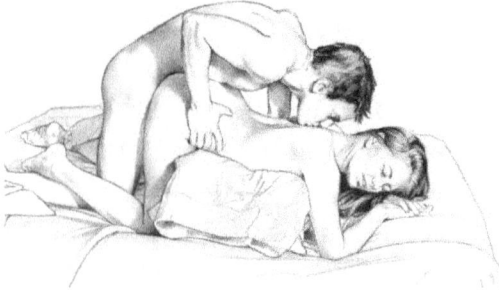

Rear Entry (Variation)

The woman can lie on her stomach with her bottom elevated.

Why she likes it: Rear-entry (or doggy-style) positions allow for very deep penetration and for manual stimulation of her clitoris by either of you.

Muscles to work: Back. You need a lot of back strength to hold your partner in position—and to hold yourself upright while thrusting forward.

Best workout: Cable Seated Row. Sit on the floor, knees bent, and grip the cable handle. Pull the handle straight back until it almost touches your waist; at the same time, pull your shoulders back and push your chest forward. Hold for a moment, then extend your arms as you return to the starting position; your shoulders and lower back should flex forward slightly.

Try for three sets of eight repetitions.

Position #10: Side By Side

You and your partner face each other while lying on your sides. It's a good position when you're both a little tired. It's also a comfortable position if you happen to have back pain or the woman is pregnant.

Side-by-side sex tends to be languorous and slow. It's easy for both partners to touch each other, and since you don't achieve deep penetration, ejaculation is often delayed.

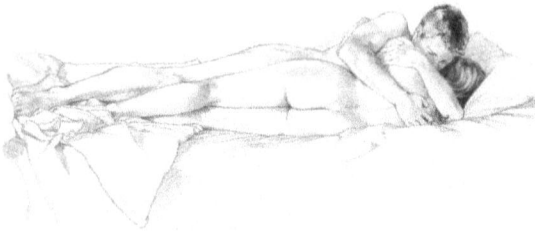

Side By Side (Variation)

You can "spoon" together, with the woman's backside nestled against you.

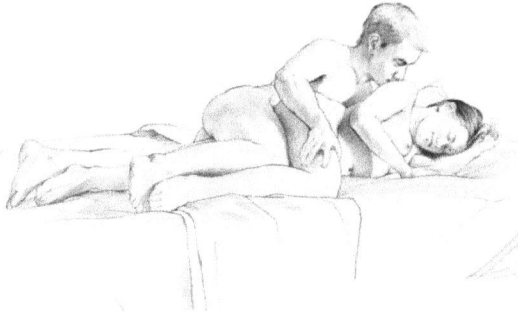

Side By Side (Variation)

You can lie on your side while the woman lies on her back, one leg between your thighs and the other leg on top.

Why she likes it: There's deep penetration in the spooning variation of the side-by-side position, it's easy for you (or her) to touch her clitoris, and it involves little vigorous thrusting — good when you're both tired.

Muscle to work: Arms. The penis tends to slip from the vagina with some frequency in this position. You need strong arms to grasp her pelvis and hold her tight.

Best workout: Incline Hammer Curl. Sit back on an incline bench with about a 45-degree angle. Hold a dumbbell in each hand, with your arms hanging at your sides. With your palms facing inward and your upper arms still, curl the dumbbells straight up, keeping your wrists locked. You can work both arms simultaneously or, if that's too difficult, alternate arms. Try for two or three sets of eight repetitions each.

Position #11: Scissors

Fun sex is a little like spirited grappling, and this move is reminiscent of something you might see in the WWF — only you both win.

She lies back on the bed with her legs bent and thighs apart. You sit between her knees and place your right leg between her thighs so your right foot is resting by her ribs under her left arm. Grasp her left leg and bring it across your body, holding it under your left arm.

Slide into her, then place your hands behind you on the bed for thrusting support. This position looks harder to achieve than it actually is, though it does require a fair amount of penile flexibility.

Why she likes it: Her clitoris will get extra stimulation as it rubs against your inner thigh during thrusting.

Why she likes it: Her clitoris will get extra stimulation as it rubs against your inner thigh during thrusting in this crisscross position.

Muscles to work: Latissimus dorsi (back muscles). A strong back makes it easier to hold her close when you're in a sitting position.

Best workout: One-Arm Dumbbell Row. Holding a dumbbell in your left hand, rest your right knee and right hand on a bench. Keep your back flat as you let the dumbbell hang down to your side so your arm lines up just in front of your shoulder.

With your left foot firmly on the floor, knee slightly bent, pull the dumbbell up and in toward your torso, raising it as high as it will go. Try for three sets of eight repetitions, then switch positions and work the other arm.

Position #12: The Pretzel

In this twisty-turny position, you recline with your right leg extended and your left leg bent. Your thighs should be apart and your torso should be off the bed about 45 degrees. She lies between your legs with her head on the bed by your right foot, weaving her right leg under your left and draping her left leg over your right as you enter her. She wraps her left arm around your right thigh, and you hold onto her waist for thrusting leverage.

Why she likes it: This pretzel position gives her easy access to massage her clitoris at a pressure she enjoys while fully reclined and comfortable.

Why she likes it: Access to her clitoris in a comfortable reclining position.

Muscles to work: Butt. You have to have strong gluts to hold this position.

Best workout: Barbell squats. Place a barbell at shoulder level on a squat rack. Grip the bar with your hands slightly more than shoulder-width apart, palms facing front. Step under the bar so that it's evenly positioned across your upper back and shoulders, not your neck. Stand up straight, with your feet hip-width apart and your knees slightly bent. Don't drop your head; keep it in line with your torso. Keeping your feet flat and torso straight, bend your knees slightly and squat down, as though sitting in a chair behind you.

Don't allow your knees to extend past your toes. Continue moving downward until your thighs are parallel to the floor. Then slowly rise to a standing position. Try for three sets of eight repetitions.

The Cardio Key To Good Sex

Athletes in full-contact sport can count on trainers to pump them full of oxygen or massage their muscles when they're on the brink of collapse. Sad to say, you can't get the same kind of help if your body suddenly gives out in the bedroom. If you like the idea of having sex all night, get used to the idea that your successes or failures are entirely up to you.
"There is nothing more embarrassing than panting and gasping when performing even the simplest moves," a 24-year-old computer support specialist in Montclair, New Jersey, told us. "I almost always get tired and have to rest," added a 52-year-old technical writer.

Here's a simple fact about the human male: He needs air. If you can't breathe, you can't make love, at least not very well. Guys who are out of shape may notice that they can't have sex and talk at the same time. They have trouble when they try to talk and have sex and change positions a few times.
There are all sorts of solutions to the kinds of sexual problems that pain men most, such as coming too quickly or not getting hard (or hard enough) when they want to. These are critical issues that deserve all the attention they get. But the foundation, the heart of good sex always comes down to air. Vigorous sex requires endurance no less than an erection. If your cardiovascular system can't cut it, neither will you.
More is involved than just kissing and breathing at the same time.
Your entire body pays the price if your heart and lungs don't work at peak capacity.

A lot of muscles come into play when you have sex. Each of them requires an abundant flow of oxygen. Limit the amount of air that comes in—and the amount of carbon dioxide that's carted out—and your strength will plummet. You'll tire easily. You'll be more likely to get debilitating, sex-stopping cramps. Blood flows to the most active muscles when you're having sex. At the same time, it continues to circulate to other body systems. If your heart isn't pounding efficiently, circulation can be significantly compromised. No, you won't die (probably), but there may be a drop in the system-wide distribution of nutrients, hormones, or other key chemicals that you need for full energy and arousal.

The landmark Massachusetts Male Aging Study, which tracked more than 600 middle-age and older men for more than 10 years, found that those who weren't active had twice the risk of impotence compared with those who took a brisk, 2-mile walk daily. Another study, this one at the New England Research Institute in Watertown, Massachusetts, reported that 31 percent of sedentary men developed impotence, compared with only 9 percent who whipped their cardiovascular systems into shape.

No matter how much you lift weights, no matter how many games of golf you play, no matter how often you sort through all the clothes that you've heaped on the now-invisible exercise bike, you have to add cardiovascular workouts into your life. You'll notice that we didn't say that you might want to think about it. You have to do it—to save your sex life as well as your life.

Stamina originates with a strong heart. The best way to strengthen the heart is with aerobic workouts: jogging, fast walking,

Can Men Fake Orgasm?

Of course. Men can fake anything. Unlike a woman's orgasm, however, men leave a little something behind. Even if your partner doesn't feel the spurt of semen — and many women don't — she's bound to notice something's amiss.

A better question to ask yourself is why you'd even want to fake an orgasm.

It's true that women (and men) sometimes fake orgasms to make their partners feel better or simply to end a sexual session for whatever reason. But this kind of deception, however well intentioned, isn't as harmless as it appears. Once you start faking anything, and once you get caught (as you will eventually), your partner will start wondering what else you lie about. Maybe, her thinking might go, you're not really attracted to her. Maybe you just pretend to like her in bed. Maybe you just pretend to like her, period.

Faking an orgasm might be a time-honored way to let your partner (and yourself) off the hook, but the possible repercussions aren't worth it.

Besides, you want to have the kind of relationship that fosters sexual openness, not deceit. Who cares if you (or your partner) don't come on occasion? It's normal. Why lie about it?

Swimming, cycling, and so on. Any time you kick your heart rate up a few notches and keep it there for half an hour or so, your heart muscle gets stronger, and your resting heart rate declines. The lungs get more efficient. Blood pressure goes down. Circulation goes up.

And on and on.

"I started running a few years ago, and my endurance went way up. Believe it or not, I have better erections now than I did when I was 30 because I don't party any more, and I work out regularly," a 45-year-old contractor in Tucson told us.

"I meet a lot more women than I used to, probably because my body image and confidence is a lot higher since I started aerobic training," says a 23-year-old student in Madison, Wisconsin.

What else can you get from regular cardiovascular training? Take a look. An increase in sex drive. A study of more than 8,000 people ages 18 to 45 found that 40 percent had increases in sexual arousal after starting a regular exercise program. One-third of them reported that they had sex more often.

Better orgasms. Researchers at the University of California report that sedentary middle-age men who started exercising for 1 hour three times weekly had better orgasms, better erections, and higher overall sexual satisfaction.

More testosterone. Aerobic workouts stimulate the body's production of testosterone, the hormone that stimulates libido and promotes erections.

Less fat. Sure, you can lose weight (or at least fat) by lifting weights, but it's a slow way to do it. Cardiovascular training puts you on the fast track. Men who walk at a leisurely, 4-mile-an-hour pace for 30 minutes will burn about 240 calories, 40 percent of which come from fat. Men who exercise more vigorously can count on burning lard at up to eight times the normal rate. Less death. Well, that's a bit of a stretch, but cardiovascular conditioning dramatically reduces the risk of heart attack, the leading cause of death in American men and women. It also cuts the risk of diabetes and colon cancer as well as high blood pressure.

A better mood. The brain churns out endorphins, opiate like brain chemicals that make you feel calm and self-assured, when you kick-start your cardiovascular system. The same chemicals that account for the runner's high can cut your risk of depression and anxiety while boosting levels of confidence and self-esteem.

"Men in some African tribes run while maintaining erections. They believe that it increases sexual stamina."

Strictly speaking, any activity performed faster than sitting in a chair is aerobic. But to produce any real benefit, there has to be substantial physical effort involved — which is why you won't see billiard players on the Olympic medal stand any time in the near future. True aerobic exercise, the kinds of workouts that tax your heart and lungs as well as your muscles, requires that you push yourself to roughly 60 to 90 percent of your maximum ability. More simply, you get an aerobic workout — also known as cardiovascular exercise — when you significantly raise your heartbeat for at least 30 minutes at a stretch.

The heart, keep in mind, is a muscle, one that gets stronger the harder you work it. As the heart gets more buffed, it starts doing the same amount of work with less effort. That's why men who work out tend to have lower resting heart rates than those who are sedentary.

The blood vessels get stronger and more elastic when you start cardiovascular training. Circulation improves throughout the body, including in the tiny blood vessels that carry blood through the penis and make erections possible. Men who start almost any kind of cardiovascular training have better bloodflow with lower blood pressure. They're less likely to develop the kinds of vascular disease that can put their sex lives out of commission. And they're more likely to have the extra energy and libido that can push their erotic pleasures to another level.

Amount. For optimal fitness, strive for a workout intensity that pushes your heart rate to 65 to 85 percent of its maximum capacity. We'll talk more about this in just a bit. In the meantime, don't let the numbers throw you. You can ballpark it by exercising with an intensity somewhere between fairly light and fairly hard. Even if you don't quite hit the target range, you'll get nearly all of the benefits, minus a percentage point or two.

Best type. This one's easy. The best aerobic workout is one that you'll actually do. Doesn't matter if it's jumping rope, swimming laps, or running hard a few steps ahead of the IRS. If you do it regularly, you'll get fit.

Consistency. Good intentions don't count. For optimal sexual (and physical and emotional) health, you need to set aside time 3 to 5 days a week. Don't expect to get the same benefits if you blow off your workouts for 3 weeks, then pound yourself into a frenzy a few days in a row. Consistency is everything — and that goes for downtime as well. When you're starting out, plan on resting for a day or two between workouts. Time off is especially important if you're running or doing other high-impact activities.

Duration. Plan on exercising for 30 to 60 minutes each time. That's a huge range, of course, but it doesn't seem to matter all that much exactly where in the range you fall. Recent studies show that men can get nearly the same health benefits from a 30-minute workout as they can from a full hour.

"A modest pucker utilizes two facial muscles, but a full-fledged, passionate kiss employs 34.
Result: A 1-minute kiss burns roughly 26 calories."

Athletes And Abstinence

When the Los Angeles Dodgers struggled early in the 2003 season, more than a few commentators floated the suggestion that their performance might improve if players took a little break from their er…, marital arrangements. None of the players actually volunteered (publicly, at least) to "give it up" for the team, and the Dodgers eventually found their footing and made the playoffs anyway.

This story is a perfect illustration of the widespread belief that athletes who have sex before competing somehow lose their edge. The myth persists even though scientific studies — yes, researchers have studied it — clearly show that having sex in no way diminishes an athlete's performance.

One study tested the grip strength of married athletes, ages 24 to 49, the morning after intercourse and the morning after abstinence.

There were no differences in grip strength on either morning. A few years later, another study looked at grip strength plus aerobic capacity and coordination. Again, sex the night before made no difference. In fact, some research suggests that athletes actually perform better when they roll in the hay before they play. down all day won't do a thing for bulk or strength. But put something heavy in your hands, and if that something is heavier than what your arms are accustomed to, you're going to get stronger.

I Love You, I Love You Not

Plenty of men having sex with one woman fill their heads with fantasies about someone else, and plenty of women do the same. Most of the time, the fantasies never make the injudicious journey from the brain to the mouth — which is a good thing, because no one wants to be called by the wrong name during moments of passion.

Is it a kind of mental infidelity to think of someone else during sex? Some women (and men) think it is. Researchers, on the other hand, say that nearly everyone does it on occasion. In most of these cases, the fantasies that occur during sex are no different from any other fantasies: They can stir your sense of eroticism without ever crossing over into reality.

There is one difference, though. In general, sharing fantasies with your partner can bring the two of you closer, but admitting that you're thinking of someone else altogether probably isn't the smartest thing to do.

Of course, all sorts of things can slip out in the heat of the moment.

Which is why more than a few wise lovers never use their partners' names during sex. Generic terms of endearment — "Oh, Baby" — work just as well and minimize the possibility of mistakes. more quickly if you vary your workouts — adding new exercises, subtracting old ones, or adding twists and refinements — about every

2 weeks. The idea is to avoid training plateaus — those dead zones where further improvement virtually stops.

Work Hard, Rest Harder

As long as men have been lifting weights, trainers have been arguing about the optimal amount of downtime. These days, the usual advice for regular guys is to lift 3 days a week — every other day, for example — and rest in between.

Men in their 20s or 30s who are already in great shape might do better with four weekly workouts. Unless you're a competitive athlete, it's highly unlikely that working out more often will provide extra benefits. If anything, it will increase your risk of injury.

Rest is important for another reason. You might think that your muscles get larger while you're lifting, but that's not what happens.

When you're at the gym, you're actually tearing down muscle fibers.

The growth phase occurs when muscles repair themselves — and that happens only after you leave the gym.

Sex Specific Strength Training

There are all sorts of reasons men lift weights. The main ones are to look better and feel better, and that's about as good as motivation gets. But in the following pages, we've done something a little different.

We talked to the country's top trainers and asked them to help us design a workout plan devoted entirely to better sex.

It's true that any weight lifting plan will go a long way toward boosting the endurance and strength that you need for good sex, but most of the exercise you get in the bedroom brings specific muscle groups into play. When you tone and strengthen these muscles, you'll find that you'll be able to have sex longer than you did before.

You'll have more energy. You'll be able to hold yourself in the most common sexual positions without fatigue.

This program is simple, yet it hits every major muscle group. It's divided into two phases. Phase 1 concentrates on building your strength base. Even if you do nothing else, it will give you strength and flexibility in places you never had them — and you (and your partner) will notice the difference within a few weeks, in bed and out.

Once you've mastered the exercises in Phase 1 — for most men, it will take several months, although some of you might be ready after

4 to 6 weeks — it's time to concentrate on specific muscles. Narrowing your focus to specific muscles will tone and sculpt your body, while at the same time improving your strength and endurance when the lights go down.

What about the "Position of Strength" exercises you see at the end of this chapter? How do those fit into the plan? The most effective way to incorporate those into your workouts is to prioritize them by starting each workout with one or two of the exercises.

As you'll see in the charts after each workout, this program requires working out 4 days a week. We've divided the charts into

"Day One" and "Day Two" workouts, each of which will be repeated twice a week, giving you time off in between.

As for equipment, you'll need either access to a gym or a nicely appointed home setup. If you don't have either, do the at-home variations.

All you'll need are some water bottles; soup cans; and heavy boxes, suitcases, sandbags, or cinder blocks. Or you might consider buying a set of dumbbells and some resistance bands. All told, they won't cost nearly as much as a home gym or gym membership.

LOWER BODY
THIS IS TH E PLACE TO START FOR B ETTE R SEX.

The most powerful muscles in your body are in the lower body. It makes sense because just about every position you find yourself in, sexual or otherwise, puts at least some strain on the legs, hips, and butt. These are also the muscles that give the power and strength to maintain and change sexual positions.

Dumbbell Lunge

START: Grab a dumbbell in each hand, with your palms facing your body, and stand with your feet hip-width apart. *FINISH:* Keeping your back straight, take a long step forward with your right leg. Bend your leg until your right thigh is parallel with the floor. Your left leg should be extended, with your knee slightly bent and almost touching the floor. Keep your right foot stationary as you straighten your right leg. Switch legs and repeat on the other side.

AT-HOME VARIATI ON: Substitute full plastic water or detergent bottles or unopened soup cans, depending on how heavy you want the weight to be. (Not pictured.)

S e x u a l G a i n s: This exercise works the butt and the front of the thigh, while putting little stress on the knee. It also strengthens the hamstrings, improves balance and posture, and supports missionary, standing, and kneeling positions.

Barbell Squat

START: Place a barbell at shoulder level on a squat rack. Grip the bar with your hands slightly more than shoulder-width apart, palms facing front. Step under the bar so that it's evenly positioned across your upper back and shoulders, not your neck. Stand up straight with your feet hip-width apart and your knees slightly bent.

Don't drop your head; keep it in line with your torso.

FINISH: Keeping your feet flat and torso straight, bend your knees slightly and squat down as though sitting in a chair behind you.

Don't allow your knees to extend past your toes. Continue moving downward until your thighs are parallel with the floor. Then slowly rise to a standing position.

AT-HOME VARIATI ON: Wall Squat

S e x u a l G a i n s: All told, the squat works more than 200 different muscles. It is the main exercise used by elite athletes for strengthening the large muscle groups — specifically, the front thigh (quadriceps), butt, and hamstrings. It also works the calves and shins, along with muscles in the shoulders, arms, and back. Great for standing and athletic positions.

Wall Squat

START: Stand with your back flat against a wall, with your feet a little wider than shoulder-width apart and your toes pointed slightly outward. Bend at your knees, keeping your weight centered over your feet.
FINISH: Lower your body as resistance, until your knees are bent
90 degrees. Straighten up slowly, concentrating on using your legs to slide up the wall. Note: Holding a heavy object — a couple of filled water bottles, for example — will give you more resistance.

UPPER BODY _
J UST AS YOU CAN'T HAVE GOOD SEX WITHOUT A STRONG
LOWER BODY, YOU CAN'T HOPE TO PERFORM AT YOUR
BEST UNLESS YOUR BACK, CHEST, AND ARMS ARE EQUALLY STRONG.

It's especially important to work the back, an area men tend to neglect. When properly strengthened and defined, the muscles give the back a visually pleasing V shape, which has the added benefit of making your waist look smaller.

Bent-over Row

START: Using an Olympic-size (45-pound) bar, wedge one end in a corner and place a 25-pound or lighter weight plate on the other end. Wrap a towel around the bar, just under the weight plate.

Straddle the bar, keeping your knees slightly bent. Your chin should be up, your chest out, stomach in, shoulders back, and back flat.

FINISH: Pull the bar to your chest, slightly arching your back and letting your elbows rise above your chest. Slowly lower the bar to arm's length, then repeat.

AT-HOME VARIATI ON: Suitcase Row (opposite)

S e x u a l G a i n s: Improves posture and form and is particularly important for vigorous sex, which tends to put a lot of pressure on the lower back.

Suitcase Row

START: Set a full suitcase (or sandbag, cinder block, railroad tie, whatever) in front of you. Stand with your legs comfortably apart, then bend over at your hips, with your knees bent and your back flat, and grab the sides of the bag.
FINISH: Use your back and biceps to pull the suitcase up to your chest, keeping it close to your body. Pause, then slowly return to the starting position.

Deadlift

START: Load a barbell and set it on the floor. Squat behind it with your feet shoulder-width apart. Grab it overhand, with your hands just outside your legs, your shoulders over or just behind the bar, your arms straight, and your back flat or slightly arched.
FINISH: Straighten up, lifting the weight to a standing position.
Push with your heels and pull the weight to your body as you stand.
Pause, then slowly return to the starting position.

AT-HOME VARIATI ON: Boxlift (opposite)
S e x u a l G a i n s: This is another superb exercise for strengthening all of the major back muscles, important for energetic sex. Be sure to do this exercise slowly. You defeat its purpose if you use momentum to finish your reps.

Box-lift

START: Set a weighted box on the floor. Stand behind the box, with your feet shoulder-width apart and your toes pointed out slightly.
Squat behind the box and grab it with a neutral grip (palms facing each other).
FINISH: Straighten up, lifting the box to a standing position. Push with your heels and pull the weight to your body as you stand.
Pause, then slowly return to the starting position.

Chin-up

START: Using an underhand grip, grab a chinup bar and place your hands shoulder-width apart. Hang from the bar with your elbows slightly bent and your ankles crossed. *FINISH:* Slowly pull yourself up until your chin is over the bar. Hold for a second or two, then slowly return to the starting position.

AT-HOME VARIATI ON: At-home variations for Chinups are not really practical. You're much better off purchasing a chinup bar.
S e x u a l G a i n s: Chinups look easy, but they require you to raise your entire body weight. They work all of the major upper- and mid-body muscles, including those in the back, chest, and arms. You need a strong back and chest to support yourself in the man-on-top position. Your partner will get a visual treat because most women are attracted to a well-developed chest.

Pushup

START: Support your body on the balls of your feet and your hands, positioning the latter slightly wider than shoulder-width apart, palms flat on the floor. Keep your eyes on the floor and your legs, back, and neck in a straight line.
FINISH: Lower your torso until your chest almost touches the floor, then slowly return to the starting position.

TO WORK YOUR CHEST MORE: Place your hands more than shoulder-width apart.
TO WORK YOUR BACK AND ARMS MORE: Place your hands together underneath your chest, thumbs and index fingers touching.
AT-HOME VARIATI ON: Not necessary. You can do Pushups anywhere, anytime.
S e x u a l G a i n s: Besides supporting you in the man-on-top position and all its variations, a strong chest and arms are required for standing positions in which you lift your partner off the floor.

Dip

START: Using the parallel bars or dip station at the gym, grab the handles with a neutral grip. Jump up and steady yourself. Start the movement with your arms straight but not locked and your body perfectly still. You can cross your legs behind you or leave them hanging straight down. The more upright you are, the harder you work your triceps. Leaning forward shifts the work to your chest and shoulders.
FINISH: Slowly lower your body until your upper arms are parallel with the floor. Push back up to the starting position. You can make the move harder by wearing a weighted dip belt or clenching a dumbbell between your ankles.

AT-HOME VARIATI ON: *Table Dip*
S e x u a l G a i n s: Dips build a strong chest, arms, and shoulders. Specifically, they work the deltoid muscles in your shoulders. They're the key to moving your arms, and they play a central role in many sexual positions, including man on top.

Surprisingly, your shoulder muscles come into play when you're performing oral sex if you're propped up on your elbows.

Table Dip

START: Position your hands shoulder-width apart on a secure table, palms down. Walk your feet out so that your knees form a 90-degree angle.
FINISH: Lower your torso until your butt is within an inch of the floor. Slowly return to the starting position.

Barbell Push Press

START: Balance a barbell on the fronts of your shoulders, with your hands shoulder-width apart and your elbows pointed up. Your knees should be slightly bent.
FINISH: Push the bar straight up until your elbows are almost locked. At the same time, rise slightly on your toes. Then slowly return to the starting position.

AT-HOME VARIATI ON: Anterior Pushup
S e x u a l G a i n s: Another great exercise for increasing the shoulder strength you need for man-on-top positions.

Anterior Pushups

START: Lie facedown on the floor, with your hands flat under your shoulders and your toes touching the floor.
FINISH: Push off the floor by extending your arms, keeping your body and legs stiff. Use only your palms and toes to hold your body up. While in the arms-extended position, round your back (this will add another inch off the floor). Finally, retract your shoulder blades, then bend your arms slowly, lowering your chest until it just barely touches the floor. Push back up again and repeat.

S e x u a l G a i n s: A variation of the Barbell Push Press, this is another great exercise for increasing shoulder strength.

ABDOMINALS _
THE ABDOMINAL MUSCLES ARE WHAT ALLOW YOU TO BEND FORWARD, BACKWARD, AND SIDE TO SIDE.

Sexually, they're the muscles that provide strong thrusting power, while at the same time allowing you to subtly alter your position and angle of thrust.

Of course, to take best advantage of your abs, you should first lose your gut. "My sex life is better now that my gut isn't in the way for either of us," says a 58-year-old retired high school principal who embarked on a weight-loss plan along with an ab-strengthening workout.

Once the gut starts to go, you'll notice that there is quite a number of muscles hidden away in there. The major ones include the transversus abdominis, which helps keep the abdominal wall tight and allows you to maintain a firm erection and control premature ejaculation. The quadratis lumborum are muscles that allow you to bend to each side. The rectus abdominis, or six-pack muscle, is the large, flat muscle running the length of the abdomen that promotes trunk flexion. The external abdominal obliques are muscles that run down the sides and front of the abdomen and allow your body to rotate and bend from side to side. The internal abdominal obliques also enable sideways movements.

Working on all these muscles has clear sexual benefits. "I've definitely noticed increased endurance, ability, and range of motion through increases in core strength," a 29-year-old computer programmer told us.

Reverse Crunch

START: Lie on your back, with your arms by your sides. Hold your legs off the floor with your knees bent at a 90-degree angle so your thighs point straight up and your lower legs point straight ahead, parallel with the floor.
FINISH: Roll your pelvis backward and raise your hips a few inches off the floor; your knees should be over your chest. Hold for a moment, then return to the starting position.

AT-HOME VARIATI ON: Not necessary. You can do most ab exercises anywhere, anytime.
S e x u a l G a i n s: This is among the best exercises for strengthening the lower abdominals, and it's among the easiest. It improves your thrusting endurance and fine-tunes subtle thrusting movements.

Twisting Oblique Crunch

START: Lie on your back, with your hands loosely touching the back of your head or neck. Bend your knees at a 90-degree angle.
FINISH: Lift your upper body off the floor and twist to the left, until your right elbow touches your left knee. Slowly return to the starting position, then repeat in the opposite direction.

AT-HOME VARIATI ON: Not necessary.
S e x u a l G a i n s: This exercise strengthens your oblique's and is good for side-by-side positions in which you enter your partner from behind and reach around her to stimulate her clitoris. You also use your oblique's anytime you attempt some of the more acrobatic positions.

Pulse-up

START: Lie with your hands flat underneath your tailbone and your legs pointed straight up toward the ceiling, perpendicular to your torso.
FINISH: Pull your navel inward and flex your gluts as you lift your hips just a few inches off the floor. Then lower your hips.

AT-HOME VARIATI ON: Not necessary.
S e x u a l G a i n s: Good for thrusting endurance and subtle thrusting movements.

WORKOUT PLAN
It's a good idea to warm up before launching into any strength training workout. Walk briskly, or ride an exercise bike for at least 5 to 10 minutes. It's also a good idea to stretch when the workout is done. See chapter 7 for a complete guide to stretching.

DAY ONE

EXERCISE	SETS	REPS	REP SPEED (SECONDS)	REST INTERVAL
Chinup (supinated)	2	10–12	2 down, 3 up	1 min.
Barbell Squat	2	10–12	2 down, 3 up	1 min.
Pushup	2	10–12	2 down, 3 up	1 min.
Deadlift	2	10–12	2 down, 3 up	2 min.
Pulse-Up	2	10–12	hold 3 at top	10 sec.
Reverse Crunch	2	10–12	2 down, 3 up	1 ½ min.

DAY TWO

EXERCISE	SETS	REPS	REP SPEED (SECONDS)	REST INTERVAL
Barbell Push Press	2	10–12	3 down, explode up	1 min.
Dumbbell Lunge	2	10–12	2 down, explode up	1 min.
Dip	2	10–12	2 down, 3 up	1 min.
Bent-Over Row	2	10–12	2 down, 3 up	2 min.
Twisting Oblique Crunch	2	10–12	2 down, 2 up	1 min.

LOWER BODY P H A S E 2

Here's where you'll start concentrating on specific muscles. As in phase 1, phase 2 has lower body, upper body, and abdominals sections, along with a biceps and triceps section.

Leg Curl

START: Lie facedown on a leg-curl machine.
FINISH: With your hips flat against the bench and your abdominal muscles tight, curl your legs behind you until your feet are about perpendicular to the bench. Pause, then lower your legs slowly to the starting position, stopping just before your knees are straight.

AT-HOME VARIATI ON: Dumbbell Leg Curl (opposite)
S e x u a l G a i n s: Strong hamstrings are essential for standing and kneeling positions.

Dumbbell Leg Curl

START: Set a dumbbell between the insteps of your feet and lie facedown on a flat bench. (This is tricky at first, and you might need someone to help you.) Grab the front or sides of the bench for support.

FINISH: Keeping your hips against the bench, curl the weight up toward your butt. Stop when your lower legs point straight up, then, without pausing, lower the weight slowly.

Calf Raise

START: Stand with the balls of your feet on the edge of a step, with your legs about 12 inches apart. Hold onto the banister or wall for stability.

FINISH: Slowly rise on your toes as far as you can go, hold for a second, then lower yourself back. To work different portions of the calves, shift your feet so that your toes point either in or out.

AT-HOME VARIATI ON: Not necessary.

S e x u a l G a i n s: Strong calves improve your endurance and strength for standing positions.

Bent-over Row

START: Stand with your feet shoulder-width apart and your knees bent 15 to 30 degrees. Keep your torso straight, with a slight arch in your back, as you lean forward at the hips. Try to get your torso close to parallel with the floor. Grab the barbell off the floor with a full overhand grip (thumbs wrapped around the bar) that's slightly wider than shoulder width. Let the bar hang at arm's length in front of you.
FINISH: Retract your shoulder blades to start pulling the bar up to the lower part of your sternum (breastbone). Pause at the top, with your chest sticking out toward the bar. Slowly return to the starting position. Try to keep your torso in the same position throughout the movement.

AT-HOME VARIATI ON: Suitcase Row (opposite)
S e x u a l G a i n s: Good for strengthening and sculpting the lower and middle portions of the back. Again, a strong back is essential for standing or kneeling positions and important for stamina during any vigorous lovemaking session.

Suitcase Row

START: Set a full suitcase (or sandbag, cinder block, railroad tie, whatever) in front of you. Stand with your legs comfortably apart, then bend over at your hips, with your knees bent and your back flat, and grab the sides of the bag. *FINISH:* Use your back and biceps to pull the suitcase up to your chest, keeping it close to your body. Pause, then slowly return to the starting position.

Lateral Pull-down

START: Sit at a lat pulldown machine. Grasp the bar with your hands about shoulder-width apart and a false overhand grip (thumbs on the same side of the bar as your fingers). Your arms should be fully extended overhead and your torso upright or leaning back slightly.
FINISH: Pull the bar straight down until it almost touches your upper chest, while squeezing your shoulder blades together. Slowly return to the starting position with your chest out, keeping full control of the bar at all times.

AT-HOME VARIATI ON: Dumbbell Pullover
S e x u a l G a i n s: This exercise works the upper part of the back, along with the shoulders and arms. It's another great exercise for man-on-top positions and standing positions in which you lift your partner off the floor.

Dumbbell Pullover

START: Lie flat on a bench. Press your head, torso, lower back, and gluts firmly against the surface. Your feet should be flat on the floor (or, if you prefer, the end of the bench). Place one hand around the handle of a dumbbell, then wrap the other over the gripping hand. Extend your arms directly above your collarbone, holding the dumbbell perpendicular to the floor.

FINISH: Keeping your back flat against the bench to elongate your lats, slowly lower the weight behind your head until your arms are in line with your ears. Pause, then pull the weight back up. Slightly bend your elbows throughout.

Sexual Gains: This exercise is a variation of the Lateral Pulldown that can be done at home. It works the upper part of the back, along with the shoulders and arms.

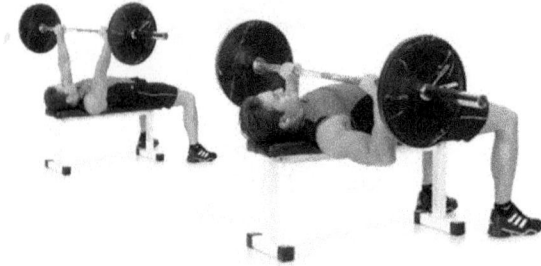

Bench Press

START: Lie flat on a bench, with your feet flat on the floor and your head positioned about eye level under the bar. Grab the barbell with a full overhand grip (thumbs wrapped around the bar), placing your hands about shoulder-width apart. Remove the bar from the uprights and hold it with straight arms over your collarbone. Pull your shoulder blades together in back.
FINISH: Lower the bar, slowly and in control, to just above your nipples. Then press it up and slightly back so it finishes above your collarbone again. Stop just short of locking your elbows, and keep your shoulder blades pulled back.

AT-HOME VARIATI ON: Use dumbbells, filled water bottles, or unopened soup cans. (Not pictured.)
Sexual Gains: The Bench Press is the main exercise for building a bigger chest, which not only looks great when you're on top but also helps hold you up.

Dumbbell Fly

START: Grab a pair of dumbbells that are lighter than those you'd use for a dumbbell bench press. Lie flat on a bench, with your feet flat on the floor and the dumbbells at arm's length above your chest.
FINISH: Maintaining a slight bend in your elbows, lower the dumbbells down and back until your upper arms are parallel with the floor and in line with your ears. Pause for a moment, then use your chest to pull the weights back to the starting position, repeating your movements in reverse. Keep your shoulder blades pulled toward each other throughout, and flex your pecs at the top of the movement.

AT-HOME VARIATI ON: Substitute unopened soup cans or filled water bottles for the dumbbells. (Not pictured.)
S e x u a l G a i n s: This is similar to the Bench Press, except you have to balance the dumbbells separately — which works your chest muscles even more.

Dumbbell Front Raise

START: Grab a dumbbell in each hand, with your arms
hanging in front of your thighs. Stand with your feet
shoulder-width apart, knees slightly bent, and lean forward
very slightly at the hips (to avoid leaning back as you lift).
FINISH: Bend your elbows slightly and raise the dumbbells
straight in front of you until your arms are parallel with the
floor. Pause, then slowly return to the starting position.

AT-HOME VARIATI ON: Substitute unopened soup cans or
filled water bottles for the dumbbells. (Not pictured.)
Sexual Gains: This exercise puts power and definition in your
shoulders—and you get an extra workout because you have to
balance the dumbbells separately. It's especially good for
when you're in the man-on-top position for an extended
period of time.

Bent-over Lateral Raise

START: Sit or stand, with your torso bent forward almost parallel with the floor and your knees slightly spread. Grab a dumbbell in each hand with your palms facing inward, elbows slightly bent.

FINISH: Slowly raise the dumbbells out to your sides, keeping your elbows bent, your back straight, and your head in a neutral position.

Squeeze your shoulder blades together at the highest point in the movement. Pause, then slowly lower the dumbbells to the starting position.

AT-HOME VARIATI ON: Substitute unopened soup cans or filled water bottles for the dumbbells. (Not pictured.)

Sexual Gains: This exercise hits the shoulders at a slightly different angle than the Front Raise, giving you that much extra man-on-top staying power. This exercise is also good for strengthening the muscles that support your neck when you're performing oral sex.

Biceps Curl

START: Holding a barbell or dumbbells with an underhand grip, stand with your feet shoulder-width apart and your arms down at your sides.

FINISH: Curl your arms up toward your shoulders. Stop and squeeze when the weight is 6 to 8 inches from your shoulders. Keep your abdomen tight, your elbows still, and your upper body straight. Pause, then lower the weight to the starting position.

AT-HOME VARIATI ON: Substitute unopened soup cans or filled water bottles for the barbell or dumbbells. (Not pictured.)

Sexual Gains: Strong biceps are necessary for lifting your partner, man-on-top positions, and rear-entry positions in which you support yourself with your arms.

Seated Dumbbell Triceps Extension

START: Sit on a bench with a 90-degree back support. With a neutral, shoulder-width grip, grab a pair of dumbbells and hold them straight up over your head with your elbows unlocked.

FINISH: Bend at the elbows as you lower the weights down to the sides of your head. Keep your upper arms in the same position and pause when your elbows are bent just past 90 degrees. Return to the starting position.

AT-HOME VARIATI ON: Use filled water bottles or unopened soup cans. (Not pictured.)

S e x u a l G a i n s: Triceps are harder to build than biceps, but you need balanced strength in your arms for maximum muscle power for lifting your partner, man-on-top, and rear-entry positions.

Triceps Kickback

START: Grab a light dumbbell in your left hand and place your right hand and knee on a bench. Plant your left foot on the floor. Bend forward at the hips so your torso is parallel with the floor. Hold the dumbbell at your side with a neutral grip, elbow pointed toward the ceiling.
FINISH: Lift the weight up and back until your arm is straight. Keep your elbow pointed toward the ceiling and the rest of your body steady. Pause for 2 full seconds, then slowly return to the starting position. Finish the set on the left side before repeating on the right.

AT-HOME VARIATI ON: Substitute unopened soup cans or filled water bottles for the dumbbells. (Not pictured.)
S e x u a l G a i n s: Besides needing strong arms to lift your partner or support yourself when you're on top, you also need them during foreplay. You can stimulate your partner manually for much longer if your arms don't get tired.

Crunch

START: Lie on your back, with your knees bent at 90 degrees or resting on a bench. Hold your hands behind your ears.
FINISH: Use your abs to curl your torso upward 4 to 6 inches, keeping your lower back firmly pressed to the floor. Pause, then slowly lower your back to the floor.

AT-HOME VARIATI ON: Not necessary.
S e x u a l G a i n s: This is the core exercise for building strong abs. Two words: Thrusting power.

Pulse-up

START: Lie with your hands underneath your tailbone and your legs pointed straight up toward the ceiling, perpendicular to your torso.
FINISH: Pull your navel inward and flex your gluts as you lift your hips just a few inches off the floor. Then lower your hips.

AT-HOME VARIATI ON: Not necessary.
Sexual Gains: Good for thrusting endurance and subtle thrusting movements.

Workout Plan Phase 2

Because this phase of the sexual fitness plan is more vigorous than the first, don't rush into it. Work your way through Phase 1 until your muscles no longer feel sore from lactic acid buildup after a workout, roughly 4 to 6 weeks after you start Phase 1.

Phase 2 is divided into push and pull days. On push days, work your chest, shoulders, and triceps — with exercises in which you push the weight. On pull days, work your back and biceps — with exercises that pull weight. Do the lower-body exercises on the second day. Plan on doing the abdominal exercises toward the end of each workout.

Don't be afraid to shuffle the exercises. Varying your workout every few weeks will hit muscles from different angles and with varying degrees of intensity, providing new stimuli ideal for optimal muscle growth. If you're satisfied with the strength and size of your arms, for example, substitute a couple of extra leg or abdominal exercises. Or maybe you're just not getting the results you need from the reverse crunch. Replace it with regular abdominal crunches.

DAY ONE (PUSH)

EXERCISE	SETS	REPS	REP SPEED (SECONDS)	REST INTERVAL
Bench Press	2	10–12	2 down, 3 up	1 min.
Bent-Over Lateral Raise	2	10–12	2 down, 3 up	1 min.
Dip	2	10–12	2 down, 3 up	1 min.
Seated Dumbbell Triceps Ext.	2	10–12	2 down, 3 up	1 min.
Dumbbell Front Raise	2	10–12	2 down, 3 up	1 min.
Triceps Kickback	2	10–12	2 down, 3 up	1 min.
Crunch	2	10–12	hold 3 at top	10 sec.
Pulse-Up	2	10–12	hold 3 at top	10 sec.
Dumbell Fly	2	10–12	2 down, 3 up	1 min.

DAY TWO (PULL)

EXERCISE	SETS	REPS	REP SPEED (SEC)	REST INTERVAL
Barbell Squat	2	10–12	2 down, explode up	1 min.
Leg Curl	2	10–12	2 down, 3 up	1 min.
Biceps Curl	2	10–12	2 down, 3 up	1 min.
Bent-Over Row 2	2	10–12	2 down, 3 up	2 min.
Lateral Pulldown	2	10–12	2 down, 3 up	1 min.
Calf Raise	2	10–12	2 down, 3 up	1 min.
Reverse Crunch	2	10–12	2 down, 2 up	1 min.
Twisting Oblique Crunch	2	10–12	2 down, 2 up	1 min.

What's Normal?

In the world of sex, not much. Normal implies some sort of ideal, or at least an average, that everyone meshes into. The concept of normal is useful for describing the behavior of large numbers of people, but it totally misses the point when you're talking about you or the guy next to you. Consider the notion of a normal number of sexual partners. Fifty years ago, the normal man was expected to sleep around a bit. The normal woman, on the other hand, was expected to sleep with one man — her husband — and one man only. Some people undoubtedly fit into these norms, but most didn't. People just lied about it more.

So forget the concept of normal. Let's look instead at what seems to be common — or not.

_ *Penis size.* Most men measure between 5 and 7 inches. Not sure how you measure up? Drop your pants, and lightly press a ruler against your abdomen alongside your erect penis.

_ *Orgasmic frequency.* In the 1940s, Dr. Alfred Kinsey interviewed
12,000 men about their sexual habits. The numbers were all over the map. Some men said it was normal for them to come four or five times a day. One gentleman disclosed that he had one orgasm in 30 years. Is there such a thing as normal? Certainly not in this arena.

_ *Sexual frequency.* About a third of heterosexual American men have sex twice a week or more. Another third has sex a few times a month. The other third has sex, at most, a few times a year. Recognize yourself in any of these categories? You might and might not because, let's face it, none of us are strictly normal. Which is why we all think, deep in our hearts, that our desires or performance somehow tilt to the abnormal side of things. And that's perfectly normal.

_ *To be won.* There's a term for guys who work this way. They're known as players, and women can smell them a mile away.

Consider what a 29-year-old graphic designer told us. "If you attempt to use a pickup line on me, no matter how smooth you think it is, you won't have a chance. Honesty and sincerity win me over all the time." All those sexual innuendos and pickup lines you've been practicing since high school? Dump them in the round file. Even if they work, and a woman agrees to go home with you, so what? You'll have a notch on your bedpost and nothing else except regrets when you find yourself, as you invariably will, alone and lonely in the middle of the night.

THROW IN SOME SURPRISES. You can never go wrong adding an element of surprise to an evening. Whether you've just met someone at a bar or are planning your first date, think of ways to show her that you're, well, a little different—and that you like her enough to invest a little energy. "I met this cute guy who bought me a few drinks, then suggested we take a walk in the park next door," a 33-year-old landscaper told us. "He didn't try to kiss me; we just walked and talked. We got to know each other better than we ever would have just sitting in the bar."

LIKE HER BEFORE YOU LOVE HER. This one is so obvious that it shouldn't need mentioning, but apparently there are a lot of men out there who equate conquest with seduction and rate their success in the world by the number of women they bed. It's not our place to judge what you do. If you want to act like a horn-dog at full moon, go for it. Just don't be surprised when you get a lot of thoroughly disgusted looks.

Women want men who like them. It's as simple as that. When you're on a first date or approaching a woman to get that first date, concentrate on her. Not her cleavage. Not what you're hoping to do later that night. Just her. What she has to say. What she does for a living. What she likes to do on Sunday mornings. The music she likes. If she has kids. Who her friends are. In other words, treat her like a person, not the trophy at the end of the race.

A man who genuinely likes the woman he's with, who shows his appreciation in a hundred different ways, and who doesn't attempt to pressure her into bed is going to get a lot further than the guy whose only intention is to score.

EXPECT NOTHING. If you doubt that women have instincts that are as finely tuned as any signal from the Hubble space telescope, just see how long you last when you start out with pure sexual vibes. She'll know what you're after, probably sooner than you do. She might let you buy her a drink. She might even enjoy flirting for a while. But you can bet she won't stick around very long. There's nothing wrong with showing sexual interest.

Women like the flirting game every bit as much as men – but only when there's something else behind it. Suppose you're sitting at a bar and decide to send a drink to a woman a few stools down. Do it because it makes you feel good to be generous. Do it because you like the way she looks and want her to notice you. Don't do it because you hope for some kind of instant payoff. Do you really expect torrid sex for the price of a measly margarita? You want the gesture to say, "Yes, I'm interested, but I'm not waiting to pounce."

FOLLOW UP. When you meet a woman for the first time, there will come the hour when you're ready for the next step. It could be backing away if you aren't (or she isn't) interested, or it could be setting up the next date. Once again, confidence goes a long way.
Come right out and ask for her phone number or e-mail address.

"Don't make me nervous by just hanging around and making me wonder if you like me or not," a 27-year-old makeup artist told us. "Be direct. Say something like, 'I'd like to spend some time with you; is there a time we can get together?'"

Programs You Can Use

There are as many effective cardiovascular training programs as there are men to invent them. Since just about any activity, done quickly, puts beneficial stress on the heart and lungs, and since every man enjoys different things, the potential permutations are endless.

But we've gone ahead and created a few basic, ready-to-go programs to make it easier: one for beginners, one for intermediate exercisers, and one for the hard-core contingent.

CARDIOVASCULAR TRAINING, PHASE 1

If you're just starting to get in shape, plan on three cardiovascular sessions a week, using a technique known as continuous training.

In other words, pick one cardiovascular exercise and do it for the full duration of your workout. You can do anything you want on different days, but make each day consistent. For example, ride a stationary bike, swim, or fast-walk for 30 to 45 minutes at a pop, making sure that your target heart rate clocks in at around 50 to 60 percent of your top capacity for at least 20 minutes of the workout.

Incidentally, don't be surprised if you notice a sudden drop in energy once you launch into your program. This is normal. Within a few weeks, you'll rebound to your previous level, then surpass it.

_ *MONDAY:* 10-minute warmup; 20 minutes of continuous exercise; 5-minute cooldown.

_ *TUESDAY:* Take the day off. And quit complaining: It wasn't that bad.

_ *WEDNESDAY:* 10-minute warmup; 20 minutes of continuous exercise; 5-minute cooldown.
_ *THURSDAY:* Rest.
_ *FRIDAY:* 10 minute warmup; 20 minutes of continuous exercise; 5-minute cooldown.
_ *SATURDAY and SUNDAY:* Rest.

CARDIOVASCULAR TRAINING, PHASE 2

This workout is designed for men who are in pretty good shape. If you've stuck with Phase 1 of the program for a few months or are otherwise active, it's time to push yourself. Here, the workout intensity is higher. You'll want to increase your target heart rate to 60 to 75 percent of your top capacity. You'll also work out more — 4 or 5 days a week instead of 3. You'll also notice that the workout is more varied. On each day of exercise, do something different: running, swimming, biking, climbing stairs, whatever. We've given exercise examples — but they're just that. Feel free to replace them with whatever aerobic activity you like best.

"Most men who start a cardiovascular training program can expect to lose at least 1 pound a week."

"Kissing can save you a fortune on dental bills.
Deep kissing stimulates the flow of saliva, which washes away food particles and lowers levels of tooth-damaging acids in the mouth."

Vitamins and Minerals

If your idea of a green vegetable is a martini olive, and you get most of your carbs from doughnuts and pretzels, your sex life may be paying the price. Yes, you need optimal levels of vitamins and minerals to protect your heart, build muscle, and strengthen bones. And yes, your risk of getting sick is a lot higher if you don't eat a fundamentally nutritious diet, preferably one that's bolstered with a daily multivitamin/ mineral supplement. But forget all that and focus on what's really important: You need a few key nutrients to maintain or improve sex drive as well as performance.

In an ideal world, we'd all eat the way nutritionists tell us to: five or more daily servings of fruits and vegetables, plenty of legumes and whole grains, minimal amounts of fat, and so on. But who has time to think about all that, especially when you're leaving the office at 7 o'clock, and your stomach essentially drags you to the nearest drive-thru lane? Even if you make a conscientious effort to eat well, your intake of key nutrients is probably on the low side, if only because so many of the foods we eat have been stripped of their natural nutritional payloads. The cells in your body, including those that make up your sexual machinery, pay the price by working at less than peak capacity.

Get in the habit of taking a multivitamin/mineral supplement. It won't miraculously boost your endurance or improve your ability to have erections, but it can give you a little edge. Obviously, every man needs a different mix of nutrients, depending on his diet and lifestyle. If you really want to do it right, see a doctor who special-izes in nutritional health to find out where you need the most help. In most cases, the nutrients below, starting with the ones men need most, are the ones that have been shown to make the biggest difference.

GET HARDER WITH ZINC.

It's probably the most important mineral for men because it's used by the body for the synthesis of testosterone. Low zinc potentially means low testosterone—which means you won't be frying any eggs on your libido until you turn things around. Men who don't get enough zinc are also likely to have uncomfortable brushes with impotence. One study at the Grand Forks Human Nutrition Research Center in North Dakota found that 11 young men who were fed low-, medium-, and high zinc diets had significantly lower levels of testosterone and lower semen volumes after eating the diet lowest in zinc for 35 days.

Your body does more with zinc than manufacture testosterone. The mineral also helps maintain an adequate sperm count and makes the penis healthier. A study by Singapore researchers compared zinc levels in the semen of 107 infertile and 103 fertile men. Those who were infertile had an average zinc level of 184 milligrams per liter compared with an average of 275 in the fertile men. The men with lower zinc levels also had lower sperm quality and motility. Zinc is also believed to sharpen your senses, especially your sense of taste and smell — and you know how important those are as part of satisfying sexual encounters. If you smoke or drink a lot of alcohol or coffee, incidentally, you'll really burn through the zinc. Make an effort to eat plenty of zinc-rich foods, such as tuna, oysters, oats, peas, and lentils. It's also a good idea to take a multi supplement that contains zinc, especially if your lifestyle habits aren't exactly on the saintly side.

TAKE THE "S EX V I TAMIN."

Vitamin E is one nutrient that deserves its racy reputation. It plays a key role in shipping blood to tis-sues throughout the body, including the penis. You need that blood for good erections. In addition, vitamin E is among the most potent antioxidants. It blocks the harmful molecules in the body that damage blood vessel walls and increase buildups of plaque, the fatty stuff that's a common cause of soft (or nonexistent) erections. Men who have diabetes are especially prone to damaged blood vessel walls, which is one reason why more than half of all men with diabetes are impotent. A study of men with and without diabetes found that those who were impotent had significantly lower vitamin E levels compared with those who did not have erectile dysfunction.

Check with your doctor first, but taking vitamin E with aspirin can significantly reduce your risk of heart disease — and the antioxidant-and-blood-thinner combination can reduce the blood-blocking arterial deposits that inhibit erections.

Vitamin E is found mainly in plant oils, along with salmon, eggs, almonds, and leafy vegetables such as spinach. It's hard to get enough from food alone, however. Of all the nutritional supplements, this is the one that your doctor is most likely to recommend.
Plan on taking 400 IU daily. This is higher than the Daily Value, so check with your doctor before taking that much.

GET ENOUGH POTASSIUM.

It balances acidity in the glands and aids in the creation of testosterone. You also need potassium for healthy nerves and muscles; men who are low in this mineral may experience diminished sexual responsiveness and coordination. Plus, potassium helps regulate blood pressure, and a study of 776 men in the Massachusetts Male Aging Study found that having high blood pressure doubled a man's risk of developing erectile dysfunction over 10 years. Another study of 54 people taking high blood pressure medication found that 81 percent of those who ate a potassium-rich diet for a year were able to cut their drug dosages significantly.

It doesn't hurt to take a daily supplement that contains potassium, but it's pretty easy to get enough as long as you eat lean meats and nuts, along with whole grains and fruits and vegetables.

PUMP SOME IRON.

Low energy in the sack? You might be low in iron, the mineral that allows red blood cells to carry oxygen to tissues in the penis and elsewhere in the body. You won't have any trouble getting enough iron as long as you eat lean meat, eggs, raisins, bananas, and green vegetables.

STRENGTHEN SPERM WITH C.

If you're thinking about starting a family, now's the time to slug down some OJ. Vitamin C increases sperm count and motility, the ability of the little squiggles to get more closely acquainted with that alluring egg. A study of 13 infertile men who were given 1,000 milligrams of vitamin C twice a day for 2 months found that the men's sperm counts more than doubled, and sperm motility improved significantly. Vitamin C improves the absorption of iron and aids in the production of the hormones that minimize stress — essential if you're hoping to have a good time when the lights go down. In one study, lab rats were confined an hour a day to increase their stress hormones. When they were fed the human equivalent of a few grams of vitamin C daily for 3 weeks, their stress hormone levels fell. Nearly every multivitamin contains adequate amounts of vitamin C. You'll get even more if you eat citrus fruits, tomatoes, peppers, onions, and other fruits and vegetables.

SIMPLIFY WITH A COMPLEX.

There are 11 different B vitamins, and you can easily drive yourself crazy figuring out how much of each you need. An easier approach is to take a B-complex supplement that includes all of these important nutrients. You need B vitamins for healthy muscles and nerves and to enhance fertility, boost circulation and the production of red blood cells, and increase energy and libido. Foods rich in B vitamins include fish, brewer's yeast, legumes, eggs, and brown rice and other whole grains.

CRANK UP TH E CALCIUM.

You already know that you need calcium for strong bones. What you may not know is that this common mineral, one of the electrolytes, is essential for the transmission of nerve signals — signals that allow the brain to efficiently communicate with the penis. Been trying some unusual positions lately? Definitely make sure you get enough calcium. Muscles need it to contract, and low levels can leave you gasping with painful, sex-stopping cramps.
Calcium also helps to keep both your blood pressure and your weight in check, and high blood pressure and obesity raise your risk of having erection problems. A survey conducted by the Harvard School of Public Health looked at the profiles of nearly 2,000 men over age 50. Those with erectile dysfunction were more likely to have high blood pressure and be overweight. In fact, men with 42- inch waistlines were nearly twice as likely to have erection problems as men whose waists measured 32 inches. A few glasses of milk or fortified juice daily will provide all the calcium you need. Other calcium-rich foods include yogurt, most breakfast cereals, and sardines (with the bones).

IMPROVE VIGOR WITH VITAMIN A.

Your body needs this nutrient to utilize the testosterone that's already present. It also strengthens testicular tissue, maintains healthy sperm levels, and inhibits the accumulation of artery-clogging gunk. You'll get plenty if you take a daily multivitamin or eat foods such as liver; eggs; and yellow and green vegetables, such as carrots, sweet potatoes, and spinach.

Herbs and Supplements

The modern pharmaceutical industry has been lobbying politicians for decades to do everything but take herbal medicines off pharmacy and health-food store shelves. You can hardly blame them. After all, all of those leaves, roots, and twigs grow free for the picking, so to speak. They're certainly a fraction of the cost of high-tech drugs. Doctors, not the most liberal bunch in town, love to talk about the inadequate testing of herbal remedies as well as the so-called dangers — though their own medical journals report that only a handful of people are harmed by herbs each year, while prescription and over-the-counter drugs are among the leading causes of injury and death in the United States.

It's true that most herbal remedies haven't been rigorously studied, but this has more to do with money than anything else. No pharmaceutical company worth its stockholders will spend millions of dollars studying natural plant compounds that can't be patented — and will never generate the gargantuan profits the industry is accustomed to. Even doctors who specialize in alternative medicine wish there were more studies to conclusively show which herbs work best, what the safest doses are, and which patients are most likely to benefit. Until those studies are done, herbal medicine will continue to represent more of a process of trial and error than one of predictable results.

Still, recent studies have shown that many herbs do appear to live up to their age-old reputations.

Indeed, the pharmaceutical industry spends a fortune sending scientists around the globe to collect and analyze plants. They have no intention of actually marketing herbs in their natural forms, of course. Instead, they use plant molecules as basic templates to create "unique" molecules that can bring astronomical profits.

"The ancient Aztecs believed that avocados conferred uncontrollable sexual desire.
Even being in proximity to one of these testicle-shaped fruits was thought to get libido raging – which is why young women, especially virgins, were required to keep a cautious distance."

"The first couple to be shown in bed together on prime-time television was Fred and Wilma Flintstone."

Summary

Use this book as a reference material, refer to it as many times as possible. Keep in mind life is good, sex should be better. Your sexual organs are made up of muscles and blood vessels, the less you use them, the weaker they get. Use It Or Lose It. You may want to try out some of our recommended books to go with your workout plans. *Green Smoothie Cleanse* to rid you off of the daily toxins from your entire system, this will help you feel lighter and healthier. It's also a good way to get in shape if you have let go of yourself for some reason. *Healthy Dinner Recipes* for those healthy late-night meals after a long day at the office or with friends and family. And *Vegetarian Recipes*, not just for vegetarians. We would love to hear from, let us and other know how this material has helped you or your partner by leaving your testimonial on your favorite bookstore today. Thank you.

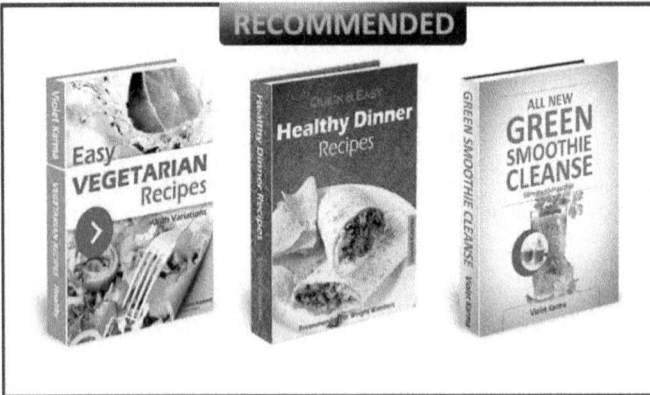

More Readings

By the same author.

Get your copy from your favorite bookstore today.
(Also available in kindle store)

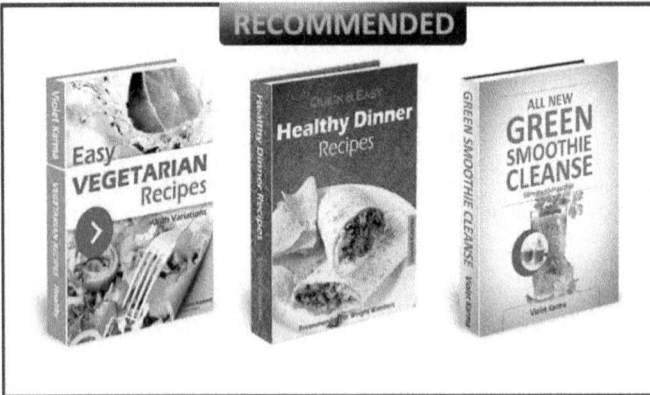

Healthy Dinner Recipes
Green Smoothie Cleanse
Vegetarian Recipes

Pump While You Hump
Penicure for Sexual Health

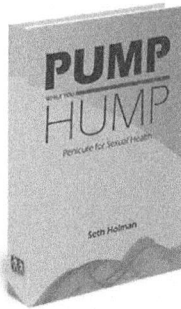

By: Seth Holman, Dr., Ph.D.

In a survey conducted by Durex Condoms, 67% of women were unhappy with their lovers' penis size. In other words, two out of every three girls prefer longer and/or bigger penis. 87% of men fail to holdback ejaculation for longer than three minutes due to weak PC muscles (puboccygeal muscles), careful use of this material will help you gain control of your PC Muscles that you could delay ejaculation and prolong the sexual pleasure between you and your partner. Weaker PC muscles can lead to weak erections, weak ejaculation, premature ejaculation and impotence. Therefore, strengthening your PC muscles helps you gain control of your prized organ.

The sole purpose of this book is to help you increase your penis size both in length and girth. Most men become insecure with their penis penile size after a series of disappointing sex, intercourse with multiple partners or after too much porn, it's merely as a result of comparison with other men. The workout, dieting and recipe suggestions in this booklet are designed to help you achieve a healthy and strong body along with even a healthier and stronger sexual drive and penis.

(Available in print in major bookstores).

The end

Lightning Source UK Ltd.
Milton Keynes UK
UKHW020634270720
367241UK00009B/614

9 781715 153779